SpringerBriefs in Electrical and Computer Engineering

Series editors

Woon-Seng Gan, School of Electrical and Electronic Engineering, Nanyang Technological University, Singapore, Singapore

C.-C. Jay Kuo, University of Southern California, Los Angeles, CA, USA

Thomas Fang Zheng, Research Institute of Information Technology, Tsinghua University, Beijing, China

Mauro Barni, Department of Information Engineering and Mathematics, University of Siena, Siena, Italy

SpringerBriefs present concise summaries of cutting-edge research and practical applications across a wide spectrum of fields. Featuring compact volumes of 50 to 125 pages, the series covers a range of content from professional to academic. Typical topics might include: timely report of state-of-the art analytical techniques, a bridge between new research results, as published in journal articles, and a contextual literature review, a snapshot of a hot or emerging topic, an in-depth case study or clinical example and a presentation of core concepts that students must understand in order to make independent contributions.

More information about this series at http://www.springer.com/series/10059

Contents

About the Authors

Hani Vahedi has received his B.Sc. and M.Sc. degrees in Power Electrical Engineering from K. N. Toosi University of Technology (KNTU), Tehran, Iran, in 2008 and Babol Noshirvani University of Technology, Babol, Iran, in 2011, respectively. He received his Ph.D. with honor from École de Technologie Superieure (ÉTS), University of Quebec, Montreal, Canada, in 2016. He is the recipient of Best PhD Thesis Award for the academic year 2016–2017 from ETS. He has published more than 60 technical papers in IEEE conferences and transactions. He has received best paper and presentation awards as well as travel assistance in numerous international conferences. He is an active member of IEEE Industrial Electronics Society (IES) and its Student and Young Professionals (S&YP) committee. He is a co-chair of special sessions and 3M video session in IES conferences and co-organizer of S&YP Forum. He also serves as an editor for International Transactions on Electrical Energy Systems, published by John Wiley & Sons, Ltd. He is the inventor of PUC5 converter and holds three US patents and transferred that technology to the industry. Currently, he is designing electric vehicle fast DC charger based on PUC5 converter at Ossiaco Inc., Montreal, Canada.

His research interests include power electronics multilevel converters topology, control and modulation techniques, power quality, active power filter, and their applications into smart grid, renewable energy conversion, UPS, battery chargers, and electric vehicles.

Mohamed Trabelsi received his B.Sc. degree in Electrical Engineering from INSAT, Tunisia, in 2006 and M.Sc. in Automated Systems and Ph.D. in Energy Systems from INSA Lyon, France, in 2006 and 2009, respectively. From October 2009 to August 2018, he has been holding different research positions at QU and TAMUQ at Qatar. In September 2018, he joined the Kuwait College of Science and Technology, where he is currently an associate professor. His research interests include systems control with applications arising in the contexts of power electronics, energy conversion, renewable energy integration, and smart grids. He has been participating as principal investigator in several collaborative research projects, resulting in more than $6 million of funding. He was recipient of the prestigious French MENRT Scholarship from the Ministry of Higher Education for postgraduate studies (2006–2009) and the Research Excellence Award for the academic years 2016–2017 and 2017–2018 in recognition of his research achievements and exceptional contributions to the Electrical and Computer Engineering Program at TAMUQ. He obtained in June 2016 a Professional Certificate of "PV System Designer and Installer" from the PV Technology Lab of the University of Cyprus. He has published more than 70 journal and conference papers and is an author of two books and one book chapter.

Chapter 1
Multilevel vs Two-Level Inverters

The amount of the energy generation and distribution systems have increased significantly in the recent years. Based on the energy statistics shown in Fig. 1.1, the world's electrical energy consumption is increasing continuously [1], which requires more power generation especially from renewable energy sources (wind and solar). From statistics [2], the total electricity consumption will be 61% higher in 2030 than in 2011. Besides, renewable energy reaches a 6% share of global energy production by 2030, up from 2% in 2011. Renewable energy resources play an important role in generating power due to green energy and low environmental impacts. However, their output is not useable by consumers and needs to be boosted and converted into a smooth AC waveform to deliver desired power to the grid with low harmonics which needs high power inverters with higher efficiency. Moreover, the industries demand higher power equipment which are more than megawatt level such as high-power AC drives which are usually connected to the medium voltage networks (2.3, 3.3, 4.16 and 6.9 kV) [3].

The output of a conventional 2-level inverter is just $+V_{dc}$ or $-V_{dc}$ from a DC capacitor with the voltage magnitude of V_{dc} that has a lot of harmonics which is vital to be filtered. Regarding these values, the switches have to suffer high amount of voltage and current if such type of inverter is used in high power applications such as mining applications, high power motor drives, PV or Wind farm energy conversion systems and etc. On the other hand, high frequency operation is also limited for high power applications due to increased power losses. Moreover, it is required to use high voltage switches which are limited by the existing technologies as shown in Fig. 1.2 [4]. One solution to overcome that limitation is using more switches and capacitors in series that can divide the voltage among the switches, which is shown in Fig. 1.3, but that increases the number of components significantly which need more DC isolators and physical space for the converter consequently [3].

To resolve the above-mentioned problems, a new technology of inverters called Multilevel Inverters (MLI) has been introduced employing combination of switches

© The Author(s), under exclusive license to Springer Nature Switzerland AG 2019
H. Vahedi, M. Trabelsi, *Single-DC-Source Multilevel Inverters*,
SpringerBriefs in Electrical and Computer Engineering,
https://doi.org/10.1007/978-3-030-15253-6_1

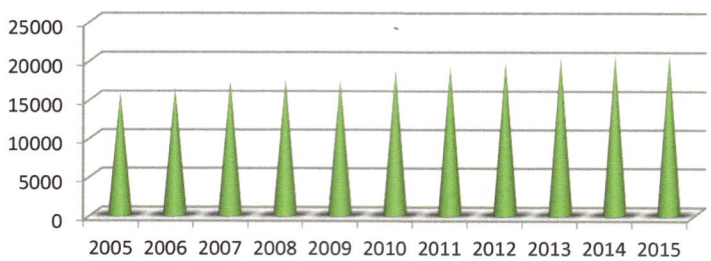

Electricity Consumption (TWh)

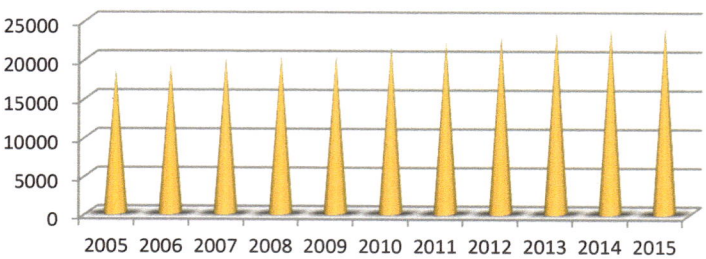

Electricity Production (TWh)

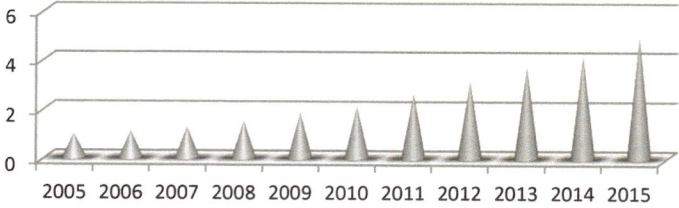

Share of Wind and Solar in Electricity Production (%)

Fig. 1.1 Electricity energy statistics in the world since 2005

and DC sources to produce various voltage levels, which is being used in medium-voltage high-power applications. Those switches are turned on and off according to a switching pattern to produce desired combination of DC voltages at the output, while the switches are not seeing the whole DC voltage and they are just blocking a part of the DC bus. As well, producing smoother waveform leads to lower harmonic which reduces the filter size and power losses remarkably. So, having less number of switches and isolated DC sources since generating high number of voltage levels at the output is always a matter of controversy where single-DC-source topologies are being considered the most suitable ones for most of the power system applications such as renewable energy conversion systems.

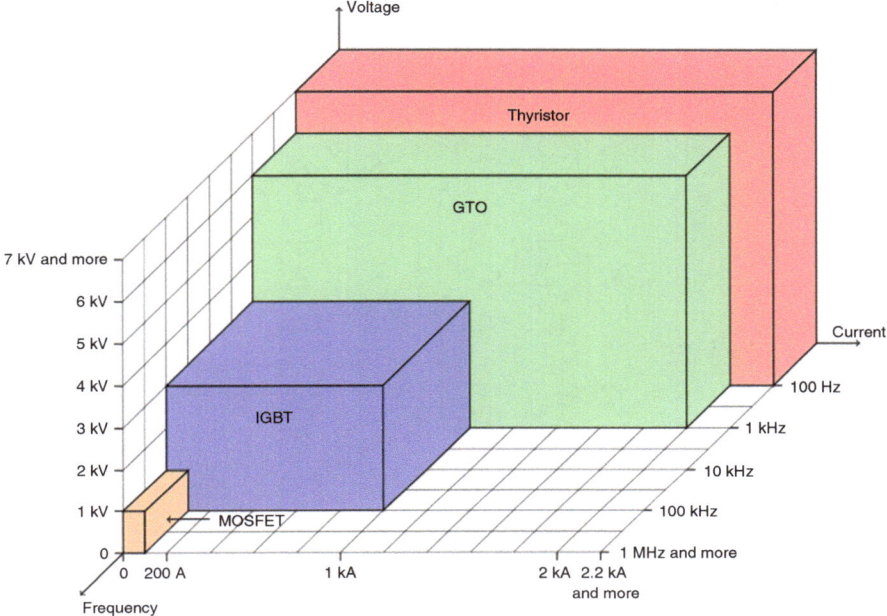

Fig. 1.2 Existing technologies of semiconductor devices in power electronics applications (photo from Wikipedia)

The MLI structure permits to generate smoother output waveform by producing different voltage levels while operating at lower switching frequency which leads to lower power losses in the power inverter and reduce the output filter size. Nowadays, the usage of such inverters has been reported up to 13.8 kV and 100 MW [3]. In such power ratings various applications for such topologies can be mentioned, e.g.: mining applications, adjustable speed drives, renewable energy conversion, utility interface devices, reactive power compensators etc. Moreover, multilevel rectifiers could be also used in high power applications such as newly emerged high power and super-fast chargers for EVs. The major weakness of conventional converters is limited power rating, high harmonic pollution and high switching frequency which prevents their usage when it comes to high power applications. Besides, in high power applications, the lower switching frequency is more desired to decrease the switching losses. Lower harmonic contents of voltage and current waveforms are also mandatory to reduce the size of output filters.

The MLI technology has been started by the concept of multilevel step wave in cascade H-Bridge converters in the late 1960s [5]. This was an attempt to present a new control method that was useful to produce and employ the stepped wave at the output of such inverters. In 1970, the Diode Clamped Converter was introduced but all these efforts were done in low power applications [6].

For medium-voltage applications, the Neutral Point Clamped (NPC) and then the Cascade H-Bridge (CHB) have been introduced in 1980s [7, 8]. In addition to these

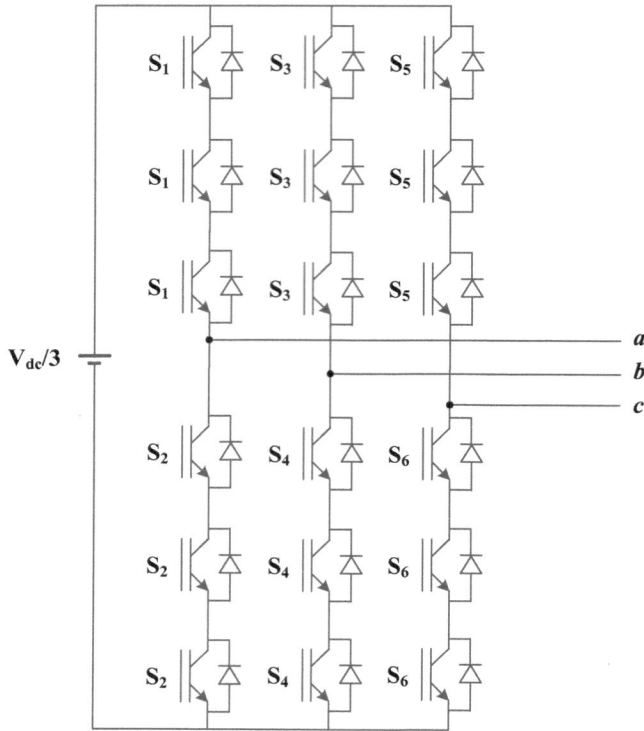

Fig. 1.3 High Power VSI with series elements (switches with same names are fired coincide)

two types, the Flying Capacitor (FC) inverter, which was introduced in 1960 as a low voltage one, has been evolved to be employed in medium-voltage and high-power industries in 1990 [9].

As an application example of such devices is medium-voltage motor drive that was begun in the middle of 1980s when the 4500 V gate turn off (GTO) thyristors were commercialized [10]. Afterwards, development of high-power switches results in manufacturing insulated gate bipolar transistor (IGBT) and gate commutated thyristor (GCT) in the late 1990s [11]. These switches have been employed in medium-voltage and high-power inverters rapidly because of their appeal characteristics, low power losses, simple gate control and snubber-less operation.

MLI consists of several semiconductor switches and DC supplies. The combination of switching actions produces various voltage levels at the output. Figure 1.4 shows the basic concept of an MLI operation. It shows the DC link and one leg of inverter in two-level, three-level and n-level configuration. The performance of semiconductor switches is shown by ideal switches. Figure 1.4a shows a conventional inverter which can produce $+V_{dc}$ or $-V_{dc}$ at the output point of 'a' with respect to the grounded neutral point, while the three-level inverter in Fig. 1.4b produces $+V_{dc}$, 0 and $-V_{dc}$ at the output and finally the n-level inverter in Fig. 1.4c generates

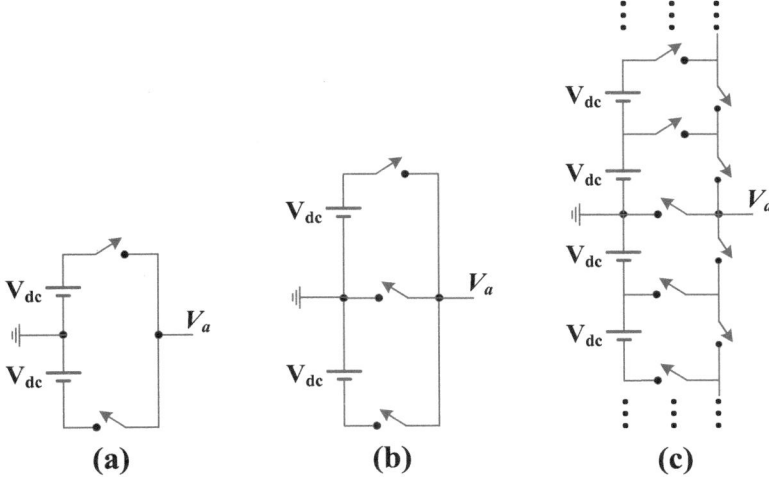

Fig. 1.4 One leg of (**a**) 2-level, (**b**) 3-level and (**c**) n-levels inverter

multilevel voltages of 0, $\pm V_{dc}$, $\pm 2V_{dc}$, ... As it is obvious from the figure, the semi-conductor switches suffer only V_{dc} or less, however the output maybe more than V_{dc}. This feature of MLI helps the industries and renewably resources to reach high power demands and applications using medium-voltage equipment.

Recently, MLI are taking attention of the researchers and industries due to their attractive features [12]. Some of the major advantages of multilevel inverters are as follows:

• Lower distortion in the output voltage due to multiple levels of output waveform;
• Lower dv/dt (voltage stress) that leads to endure the reduced voltage by switches;
• Lower common mode voltage which is helpful in motor drives;
• Lower switching frequency results in lower switching losses.

Different types of MLI have been proposed and built, which are mostly for medium-voltage and high-power applications because of the fact that a single power switch cannot be connected to a medium-voltage grid directly.

References

1. Enerdata. Global Energy Statistical Yearbook 2015 [Online]. Available: http://yearbook.enerd-ata.net/
2. British-Petroleum, "BP World Energy Outlook 2030," January, 2013.
3. J. Rodríguez, S. Bernet, B. Wu, J. O. Pontt, and S. Kouro, "Multilevel voltage-source-converter topologies for industrial medium-voltage drives," *IEEE Trans. Ind. Electron.*, vol. 54, no. 6, pp. 2930–2945, 2007.

4. Wikipedia. (2012). *Power Semiconductor Device*. Available: http://en.wikipedia.org/wiki/ Power_semiconductor_device
5. W. McMurray, "Fast response stepped-wave switching power converter circuit," US Patent 3581212, 1971.
6. R. H. Baker, "High-voltage converter circuit," US Patent 4203151, 1980.
7. R. H. Baker, "Bridge converter circuit," 4270163, 1981.
8. A. Nabae, I. Takahashi, and H. Akagi, "A new neutral-point-clamped PWM inverter," *IEEE Trans. Ind. Applications,* no. 5, pp. 518–523, 1981.
9. T. Meynard and H. Foch, "Dispositif électronique de conversion d'énergie électrique," France Patent, 1991.
10. B. Wu, *High-power converters and AC drives*: Wiley-IEEE Press, 2006.
11. P. K. Steimer, H. E. Gruning, J. Werninger, E. Carroll, S. Klaka, and S. Linder, "IGCT-a new emerging technology for high power, low cost inverters," in *Industry Applications Conference, 1997. Thirty-Second IAS Annual Meeting, IAS'97., Conference Record of the 1997 IEEE,* 1997, pp. 1592–1599.
12. S. Kouro, M. Malinowski, K. Gopakumar, J. Pou, L. G. Franquelo, B. Wu, *et al.,* "Recent advances and industrial applications of multilevel converters," *IEEE Trans. Ind. Electron.,* vol. 57, no. 8, pp. 2553–2580, 2010.

Chapter 2
Multi DC Source Inverters, Pros and Cons

There are some certain differences between single-DC-source and multiple-DC-source MLI topologies which limit the application of multiple-DC-source ones in power systems. Such limitations are discussed based on the topology and associated controller approach, separately.

Saying from topology point of view, multiple-DC-source inverters need more than one isolated DC supply [1, 2]. Therefore, they will not be cost-effective and small size because an isolated DC supply is made up of a transformer and a diode bridge, or it could be a battery or PV panel as shown in Fig. 2.1 Consequently, multiple-DC-source topologies have at least one more supply than a single-DC-source one, which means undesired additional size and cost. The most popular MLI with multiple-DC-source is the CHB as shown in Fig. 2.2a, which is used in very high-power motor drives currently. The main advantages of CHB are the modularity and identical voltage rating of switches due to using equal DC sources.

Many other multiple-DC-source topologies have been published but they suffer from unequal voltage rating of DC supplies and switches which is the main reason of not getting attraction from industries. A fair comparison between CHB and all those multiple-DC-source MLIs considering equal voltage rating of components reveals the fact that CHB is still the best one with optimum number of components due to its identical voltage rating in each cell and modularity. A major concern with multiple-DC-source configurations is the power sharing among feeders. An unbalanced power sharing causes undesirable power losses and malfunctioning [3, 4].

Considering the control algorithms, a cascaded control consisting of a current and a voltage loop is required for most of the power converters applications [5]. Focusing on grid-connected ones, such as PV systems, the DC bus should be also regulated by injecting the voltage error into the current reference [6]. Such scheme as shown in Fig. 2.2b, requires numerous voltage loops and consequently voltage PI regulators for multiple-DC-source topologies. Moreover, implementing a power-balancing unit is inevitable to share the appropriate amount of power between

© The Author(s), under exclusive license to Springer Nature Switzerland AG 2019
H. Vahedi, M. Trabelsi, *Single-DC-Source Multilevel Inverters*,
SpringerBriefs in Electrical and Computer Engineering,
https://doi.org/10.1007/978-3-030-15253-6_2

Fig. 2.1 Real examples of isolated DC sources

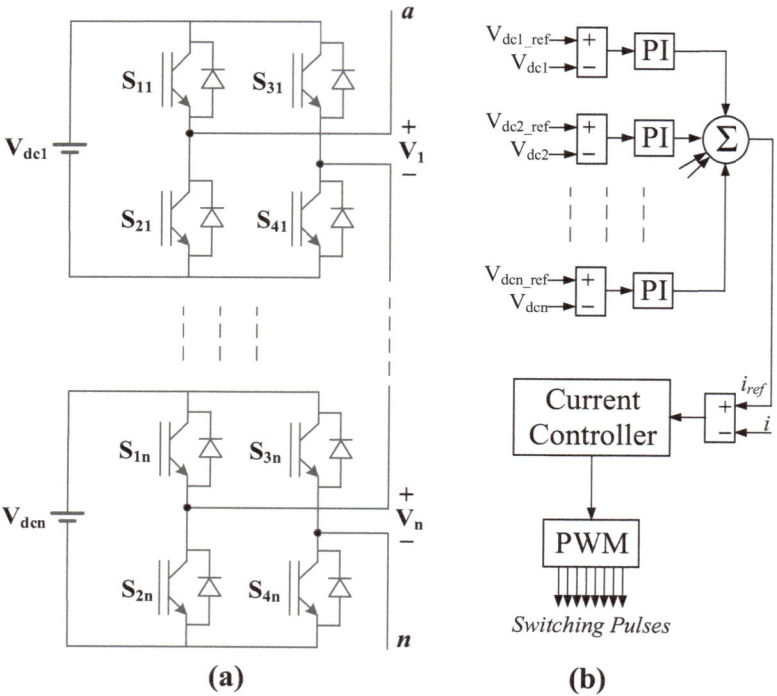

Fig. 2.2 Multiple-DC-source inverters (**a**) CHB (**b**) Corresponding controller

sources [7, 8]. As well, injecting all voltage errors into only one single current reference cannot ensure proper distributing of the active power among DC links to keep the capacitors voltages fixed. This issue also exists in single-DC-source inverters in which there is no redundant switching states to balance the auxiliary capacitors voltages. For instance, the two DC capacitors in Neutral-Point Clamped (NPC) and T3 inverters can be controlled by redundant switching states only in 3-phase configuration; otherwise, they have to use external controllers for 1-phase applications [9, 10]. The same concern exists for Active Neutral-Point Clamped (ANPC) and Flying Capacitors (FC) inverters except their auxiliary capacitors [11].

As a conclusion, a single-DC-source inverter is desired in which the auxiliary capacitors are controlled through switching states without adding extra linear/non-linear regulators and complexity to the system. Such topology can be installed in all power system applications where the 2-level ones are already operating. Therefore, the input DC side and output AC side do not require to be modified. Moreover, the controller remains the same since only a single DC link should be regulated and the error signal goes into the reference current. However, the modulation block should be replaced by a multilevel switching technique with integrated voltage balancing using redundant switching states.

References

1. Y. Hinago and H. Koizumi, "A single-phase multilevel inverter using switched series/parallel dc voltage sources," *IEEE Trans. Ind. Electron.,* vol. 57, no. 8, pp. 2643–2650, 2010.
2. K. K. Gupta and S. Jain, "A novel multilevel inverter based on switched DC sources," *IEEE Trans. Ind. Electron.,* vol. 61, no. 7, pp. 3269–3278, 2014.
3. E. Samadaei, S. A. Gholamian, A. Sheikholeslami, and J. Adabi, "An envelope type (E-Type) module: asymmetric multilevel inverters with reduced components," *IEEE Trans. Ind. Electron.,* vol. 63, no. 11, pp. 7148–7156, 2016.
4. M. Malinowski, K. Gopakumar, J. Rodriguez, and M. A. Perez, "A survey on cascaded multilevel inverters," *IEEE Trans. Ind. Electron.,* vol. 57, no. 7, pp. 2197–2206, 2010.
5. H. Vahedi, A. Shojaei, A. Chandra, and K. Al-Haddad, "Five-Level Reduced-Switch-Count Boost PFC Rectifier with Multicarrier PWM," *IEEE Trans. Ind. Applications,* vol. 52, no. 5, pp. 4201–4207, 2016.
6. Y. Liu, B. Ge, H. Abu-Rub, and F. Z. Peng, "An effective control method for three-phase quasi-Z-source cascaded multilevel inverter based grid-tie photovoltaic power system," *IEEE Trans. Ind. Electron.,* vol. 61, no. 12, pp. 6794–6802, 2014.
7. S. Vazquez, J. Leon, J. M. Carrasco, L. G. Franquelo, E. Galvan, M. Reyes, *et al.*, "Analysis of the power balance in the cells of a multilevel cascaded H-bridge converter," *IEEE Trans. Ind. Electron.,* vol. 57, no. 7, pp. 2287–2296, 2010.
8. Z. Liu, B. Liu, S. Duan, and Y. Kang, "A novel DC capacitor voltage balance control method for cascade multilevel STATCOM," *IEEE Trans. Power Electron.,* vol. 27, no. 1, pp. 14–27, 2012.
9. Z. Shu, X. He, Z. Wang, D. Qiu, and Y. Jing, "Voltage Balancing Approaches for Diode-Clamped Multilevel Converters Using Auxiliary Capacitor-Based Circuit," 2013.
10. M. Schweizer and J. W. Kolar, "Design and implementation of a highly efficient three-level T-type converter for low-voltage applications," *IEEE Trans. Power Electron.,* vol. 28, no. 2, pp. 899–907, 2013.
11. H. Wang, L. Kou, Y.-F. Liu, and P. C. Sen, "A New Six-Switch Five-Level Active Neutral Point Clamped Inverter for PV Applications," *IEEE Trans. Power Electron.,* vol. 32, no. 9, pp. 6700–6715, 2017.

Chapter 3
Single-DC-Source Multilevel Inverters

In this chapter, some Single-DC-Source MLI topologies and their voltage balancing techniques are surveyed.

3.1 Topologies

Neutral-Point Clamped (NPC), T3, ANPC, FC, and Modular Multilevel Converter (MMC) are the most attractive single-DC-source topologies in which flying capacitors (auxiliary ones) are used to produce more voltage levels at the output.

3.1.1 Neutral-Point Clamped and T3

One of the mostly used single-DC-Source MLI topologies is the 3-level NPC (Fig. 3.1a) [1] and the modified version of NPC which is called T3 shown in Fig. 3.1b [2]. Indeed, the clamping diodes of the NPC are replaced by bidirectional switches in T3. Those configurations can generate 5-level voltage waveform at the output.

The multilevel NPC and T3 inverters are widely used in many applications, such as medium voltage drive, marine, and mining activities. This is due to its good static and dynamic performance compared to the 2-level structure. However, the main drawback of those topologies is the midpoint balancing challenge, which is mostly realized by a capacitive divider bridge. The voltage potential of the mid-point might have large ripples. To solve these problems, there are several solutions such as the use of space vector modulation, linear/nonlinear controllers or the implementation of an external circuit dedicated to the balancing of the capacitor voltages. But these solutions become more complex and add additional costs to the converter.

© The Author(s), under exclusive license to Springer Nature Switzerland AG 2019
H. Vahedi, M. Trabelsi, *Single-DC-Source Multilevel Inverters*,
SpringerBriefs in Electrical and Computer Engineering,
https://doi.org/10.1007/978-3-030-15253-6_3

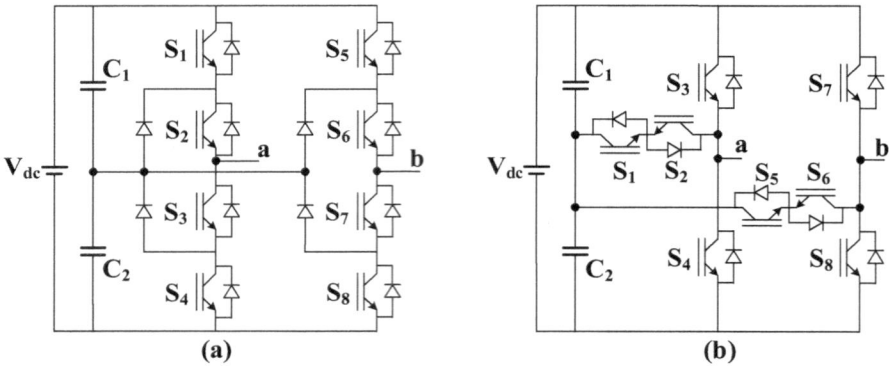

Fig. 3.1 NPC based topologies: (**a**) standard NPC (**b**) T3

Fig. 3.2 ANPC inverter
topology

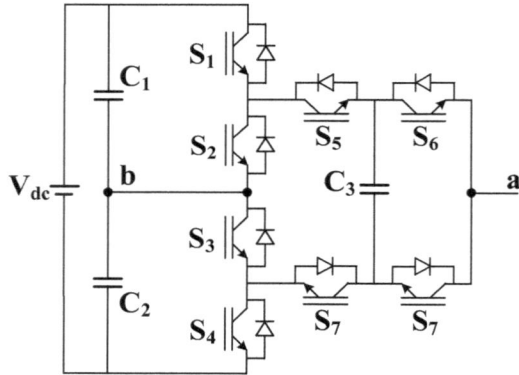

3.1.2 Active NPC

ANPC inverter has been developed as a remedy to NPC issues in power losses man-
agement [3]. The structure is shown in Fig. 3.2. It has the same voltage balancing
challenge as NPC and T3. Therefore, a modulation technique, external controller or
circuit should be added to achieve a reliable operation.

3.1.3 Flying Capacitors Converter

Recently, FC converters (FCC) have drawn more attention from industry and aca-
demia (Fig. 3.3) owing to their merits such as natural voltage balancing feature,
transformer-less operation, and equally distribution of switching stress among
semiconductor power switches. FCC are comprised of switching power cells

Fig. 3.3 Single-phase
FCC circuit

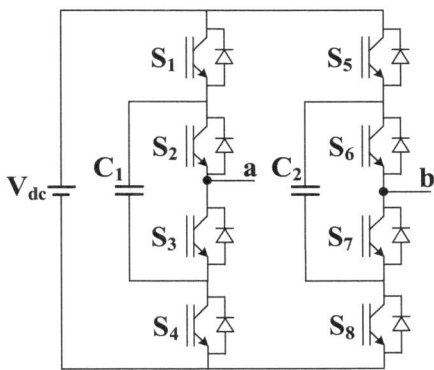

connected in series to form the converter phase leg. Each switching power cell is realized by two low−/medium-voltage transistors possessing a complementary state with respect to each other and one flying capacitor [4, 5]. This implies that using these series-connected switching power cell, it is possible to achieve higher voltage/power ranges (multilevel output voltage). To avoid the short-circuit of the voltage source, the switches S1 and S4, or S2 and S3 must be controlled in a complementary manner. Conventionally, phase-shifted pulse-width modulation (PS-PWM) techniques [6, 7] are used to control FCCs.

3.1.4 Other Single-DC-Source Multilevel Inverter Topologies

Most of the single-dc-source multilevel topologies are derived from FCC such as: stacked multicell converter (SMC) [8, 9], double flying capacitor (DFC) [10], improved DFC [11], and cascaded connection of the modified FCC. Some step-up MLIs have been presented in [12, 13]. They are developed by numerous switches and capacitors and can increase the maximum voltage amplitude by connecting different charged capacitors in series. Another topology has been presented in [14, 15] that connects the H-bridge cells in a way to achieve a 7-level MLI using a single DC source. Another innovating topology is shown in [16] which is similar to the T3 but with 3 DC capacitors in DC link to split the voltage levels into 7 identical ones and generate a 7-level voltage waveform at the output. A 9-level single-dc-source topology has been reported in [17] which is based on switched-capacitor cascaded H-bridges. Another 9-level topology has been developed based on ANPC structure as analyzed in [18]. Some single-DC-source MLIs with high number of voltage levels have been also reported in [19, 20{Narimani, 2015 #252, 21}]. As well, the packed U-cell (PUC) inverter has been developed as 5-level and 7-level topologies while using a single-dc-source [22–30]. It will be thoroughly analyzed at the end of this book.

As a general comment, it should be noted that generating high number of levels while using a single-DC-source, requires more DC capacitors which complicates the voltage balancing process. As a result, such configurations may have a limited range of operation in terms of power rating, power factor, modulation index, etc. However, as analyzed in this book, the only solution would be the enough number of redundant switching states to balance the capacitors voltages efficiently through the switching pattern.

3.1.5 Modular Multilevel Converter

The MMC is a very promising and attractive converter used especially in high-voltage direct current (HVDC) transmission system applications [31]. Recently, the MMC became the center of interest for many researches, whereas many configurations and control methods have been developed. Single-phase and three-phase topologies are both used [32, 33]. Figure 3.4 shows a single phase 3-level MMC circuit, where $SM_{i\{1-4\}}$ denotes the cascaded connection of multiple submodules. Note that there are two arms (upper and lower) in each leg, which are connected in series through two identical inductors L. The MMC is also used as a converter for dc or ac motor-drive applications [34]. For ac motor drives, the MMC is called

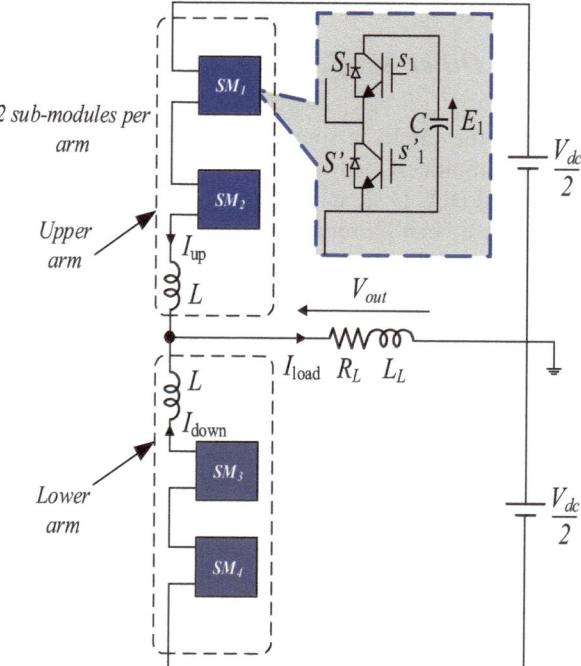

Fig. 3.4 Schematic of a 3-level MMC

modular multilevel inverter (MMI). The main advantage of this structure is that the degrees of freedom increases (redundancy in generating voltage levels) when the number of sub-modules increases, which makes it a fault-tolerant topology.

3.2 Capacitor Voltage Balancing Techniques

The main challenge for most of the above-mentioned topologies is the voltage balancing of auxiliary capacitors, which could be performed through redundant switching states, external controllers or additional circuits. It is worth mentioning that using redundant switching states is the most convenient technique to balance the capacitors voltages as already explained in Chap. 2. For instance, authors in [35, 36] proposed a voltage balancing method for FCCs using Phase Disposition Pulse Width Modulation (PD-PWM) strategy. The same technique was used in [37] for balancing the capacitors voltages of a 7-level 3×2 SMC. In [38], authors proposed an optimal switching-based voltage balancing method for a similar SMC structure, where only optimal transitions between consecutive voltage levels are used in order to reduce the switching transitions of the power devices. Model Predictive Control (MPC) with its different forms has been also extensively used as an alternative to the standard PI controllers to control FCCs and MMCs due to its fast dynamic response and ability to include constraints with different natures [26–28, 30, 39–45]. In [46], authors proposed a weighted MPC approach for a single-phase 4-cells MMC topology. The proposed controller was designed to control the load current while keeping minimum circulating current and balancing the capacitor voltages. At each sampling time, the proposed controller predicts the next switching pattern that assures best tracking of the reference variables.

References

1. A. Nabae, I. Takahashi, and H. Akagi, "A new neutral-point-clamped PWM inverter," *IEEE Trans. Ind. Applications,* no. 5, pp. 518–523, 1981.
2. M. Schweizer and J. W. Kolar, "Design and implementation of a highly efficient three-level T-type converter for low-voltage applications," *IEEE Trans. Power Electron.,* vol. 28, no. 2, pp. 899–907, 2013.
3. H. Wang, L. Kou, Y.-F. Liu, and P. C. Sen, "A New Six-Switch Five-Level Active Neutral Point Clamped Inverter for PV Applications," *IEEE Trans. Power Electron.,* vol. 32, no. 9, pp. 6700–6715, 2017.
4. M. F. Escalante, J. C. Vannier, and A. Arzandé, "Flying capacitor multilevel inverters and DTC motor drive applications," *IEEE Trans. Ind. Electron.,* vol. 49, no. 4, pp. 809–815, 2002.
5. J. Mathew, P. Rajeevan, K. Mathew, N. A. Azeez, and K. Gopakumar, "A Multilevel inverter scheme with dodecagonal voltage space vectors based on flying capacitor topology for induction motor drives with open-end winding configuration," 2013.
6. B. P. McGrath and D. G. Holmes, "Multicarrier PWM strategies for multilevel inverters," *IEEE Trans. Ind. Electron.,* vol. 49, no. 4, pp. 858–867, 2002.

7. A. Shukla, A. Ghosh, and A. Joshi, "Natural balancing of flying capacitor voltages in multicell inverter under PD carrier-based PWM," *IEEE Trans. Power Electron.*, vol. 26, no. 6, pp. 1682–1693, 2011.
8. M. Ben Smida and F. Ben Ammar, "Modeling and DBC-PSC-PWM control of a three-phase flying-capacitor stacked multilevel voltage source inverter," *IEEE Trans. Ind. Electron.*, vol. 57, no. 7, pp. 2231–2239, 2010.
9. A. K. Sadigh, V. Dargahi, M. A. Pahlavani, and A. Shoulaie, "Elimination of one dc voltage source in stacked multicell converters," *IET Power Electron.*, vol. 5, no. 6, pp. 644–658, 2012.
10. A. K. Sadigh, S. H. Hosseini, M. Sabahi, and G. B. Gharehpetian, "Double flying capacitor multicell converter based on modified phase-shifted pulsewidth modulation," *IEEE Trans. Power Electron.*, vol. 25, no. 6, pp. 1517–1526, 2010.
11. V. Dargahi, A. K. Sadigh, M. Abarzadeh, M. R. A. Pahlavani, and A. Shoulaie, "Flying capacitors reduction in an improved double flying capacitor multicell converter controlled by a modified modulation method," *IEEE Trans. Power Electron.*, vol. 27, no. 9, pp. 3875–3887, 2012.
12. H. Vahedi, K. Al-Haddad, Y. Ounejjar, and K. Addoweesh, "Crossover Switches Cell (CSC): A New Multilevel Inverter Topology with Maximum Voltage Levels and Minimum DC Sources," in *IECON 2013-39th Annual Conference on IEEE Industrial Electronics Society*, Austria, 2013, pp. 54–59.
13. A. Taghvaie, J. Adabi, and M. Rezanejad, "A self-balanced step-up multilevel inverter based on switched-capacitor structure," *IEEE Trans. Power Electron.*, vol. 33, no. 1, pp. 199–209, 2018.
14. S. S. Lee, "A Single-Phase Single-Source 7-Level Inverter with Triple Voltage Boosting Gain," *IEEE Access*, 2018.
15. H. Vahedi, M. Sharifzadeh, K. Al-Haddad, and B. M. Wilamowski, "Single-DC-source 7-level CHB inverter with multicarrier level-shifted PWM," in *IECON 2015-41st Annual Conference of the IEEE Industrial Electronics Society*, Japan, 2015, pp. 4328–4333.
16. J.-S. Choi and F.-s. Kang, "Seven-level PWM inverter employing series-connected capacitors paralleled to a single DC voltage source," *IEEE Trans. Ind. Electron.*, vol. 62, no. 6, pp. 3448–3459, 2015.
17. M. Saeedian, E. Pouresmaeil, E. Samadaei, E. Manuel Godinho Rodrigues, R. Godina, and M. Marzband, "An Innovative Dual-Boost Nine-Level Inverter with Low-Voltage Rating Switches," *Energies*, vol. 12, no. 2, p. 207, 2019.
18. K. Wang, Z. Zheng, D. Wei, B. Fan, and Y. Li, "Topology and Capacitor Voltage Balancing Control of a Symmetrical Hybrid Nine-Level Inverter for High-Speed Motor Drives," *IEEE Trans. Ind. Applications*, vol. 53, no. 6, pp. 5563–5572, 2017.
19. K. Gopakumar, M. Boby, A. K. Yadav, L. G. Franquelo, and S. S. Williamson, "Multilevel 24-Sided Polygonal Voltage-Space-Vector Structure Generation for an IM Drive Using a Single DC Source," *IEEE Trans. Ind. Electron.*, vol. 66, no. 2, pp. 1023–1031, 2019.
20. A. Kshirsagar, R. S. Kaarthik, A. Rahul, K. Gopakumar, L. Umanand, S. K. Biswas, *et al.*, "17-level inverter with low component count for open-end induction motor drives," *IET Power Electronics*, vol. 11, no. 5, pp. 922–929, 2017.
21. M. Narimani, B. Wu, and N. R. Zargari, "A new five-level nested neutral point clamped (NNPC) voltage source converter," in *Applied Power Electronics Conference and Exposition (APEC), 2017 IEEE*, 2017, pp. 2554–2558.
22. Y. Ounejjar, K. Al-Haddad, and L. A. Grégoire, "Packed U cells multilevel converter topology: theoretical study and experimental validation," *IEEE Trans. Ind. Electron.*, vol. 58, no. 4, pp. 1294–1306, 2011.
23. Y. Ounejjar, K. Al-Haddad, and L. A. Dessaint, "A Novel Six-Band Hysteresis Control for the Packed U Cells Seven-Level Converter: Experimental Validation," *IEEE Trans. Ind. Electron.*, vol. 59, no. 10, pp. 3808–3816, 2012.
24. H. Vahedi and K. Al-Haddad, "Real-Time Implementation of a Seven-Level Packed U-Cell Inverter with a Low-Switching-Frequency Voltage Regulator," *IEEE Trans. Power Electron.*, vol. 31, no. 8, pp. 5967–5973, 2016.

25. H. Vahedi, P. Labbe, and K. Al-Haddad, "Sensor-Less Five-Level Packed U-Cell (PUC5) Inverter Operating in Stand-Alone and Grid-Connected Modes," *IEEE Trans. Ind. Informat.*, vol. 12, no. 1, pp. 361–370, 2016.
26. J. Metri, H. Vahedi, H. Kanaan, and K. Al-Haddad, "Real-Time Implementation of Model Predictive Control on 7-Level Packed U-Cell Inverter," *IEEE Trans. Ind. Electron.*, vol. 63, no. 7, pp. 4180–4186, 2016.
27. M. Trabelsi, S. Bayhan, K. A. Ghazi, H. Abu-Rub, and L. Ben-Brahim, "Finite-control-set model predictive control for grid-connected packed-U-cells multilevel inverter," *IEEE Trans. Ind. Electron.*, vol. 63, no. 11, pp. 7286–7295, 2016.
28. M. Trabelsi, S. Bayhan, M. Metry, H. Abu-Rub, L. Ben-Brahim, and R. Balog, "An effective Model Predictive Control for grid connected Packed U Cells multilevel inverter," in *Power and Energy Conference at Illinois (PECI)*, 2016, pp. 1–6.
29. S. Xiao, M. Metry, M. Trabelsi, R. S. Balog, and H. Abu-Rub, "A Model Predictive Control technique for utility-scale grid connected battery systems using packed U cells multilevel inverter," in *IECON 2016-42nd Annual Conference of the IEEE Industrial Electronics Society*, 2016, pp. 5953–5958.
30. F. Sebaaly, H. Vahedi, H. Kanaan, and K. Al-Haddad, "Experimental Design of Fixed Switching Frequency Model Predictive Control for Sensorless Five-Level Packed U-Cell Inverter," *IEEE Trans. Ind. Electron.*, vol. 66, no. 5, pp. 3427–3434, 2019.
31. A. Nami, J. Liang, F. Dijkhuizen, and G. D. Demetriades, "Modular multilevel converters for HVDC applications: Review on converter cells and functionalities," *IEEE Trans. Power Electron.*, vol. 30, no. 1, pp. 18–36, 2015.
32. H. Nademi, A. Das, R. Burgos, and L. E. Norum, "A new circuit performance of modular multilevel inverter suitable for photovoltaic conversion plants," *IEEE Journal Emerg. and Select. Topics in Power Electron.*, vol. 4, no. 2, pp. 393–404, 2016.
33. I. Gowaid, G. Adam, A. Massoud, S. Ahmed, and B. Williams, "Hybrid and Modular Multilevel Converter Designs for Isolated HVDC-DC Converters," *IEEE Journal Emerg. and Select. Topics in Power Electron.*, vol. PP, no. 99, p. 1, 2017.
34. M. A. Perez, J. Rodriguez, E. J. Fuentes, and F. Kammerer, "Predictive control of AC–AC modular multilevel converters," *IEEE Trans. Ind. Electron.*, vol. 59, no. 7, pp. 2832–2839, 2012.
35. Z. Lim, A. I. Maswood, and G. H. Ooi, "Modular-cell inverter employing reduced flying capacitors with hybrid phase-shifted carrier phase-disposition PWM," *IEEE Trans. Ind. Electron.*, vol. 62, no. 7, pp. 4086–4095, 2015.
36. A. M. Ghias, J. Pou, V. G. Agelidis, and M. Ciobotaru, "Voltage balancing method for a flying capacitor multilevel converter using phase disposition PWM," *IEEE Trans. Ind. Electron.*, vol. 61, no. 12, pp. 6538–6546, 2014.
37. A. M. Ghias, J. Pou, and V. G. Agelidis, "Voltage-balancing method for stacked multicell converters using phase-disposition PWM," *IEEE Trans. Ind. Electron.*, vol. 62, no. 7, pp. 4001–4010, 2015.
38. T. Premkoumar, M. Rashmi, A. Suresh, and D. R. Warrier, "Optimal voltage balancing method for reduced switching power losses in stacked multicell converters," in *International Conference on Information Communication and Embedded Systems (ICICES)*, 2017, pp. 1–5.
39. S. Kouro, P. Cortés, R. Vargas, U. Ammann, and J. Rodríguez, "Model predictive control—A simple and powerful method to control power converters," *IEEE Trans. Ind. Electron.*, vol. 56, no. 6, pp. 1826–1838, 2009.
40. V. Monteiro, J. C. Ferreira, A. A. N. Melendez, and J. L. Afonso, "Model Predictive Control Applied to an Improved Five-Level Bidirectional Converter," *IEEE Trans. Ind. Electron.*, vol. PP, no. 99, pp. 1–1, 2016.
41. S. Vazquez, R. Aguilera, P. Acuna, J. Pou, J. Leon, L. Franquelo, *et al.*, "Model Predictive Control for Single-Phase NPC Converters Based on Optimal Switching Sequences," *IEEE Trans. Ind. Electron.*, vol. PP, no. 99, pp. 1–1, 2016.

42. M. Trabelsi, L. Ben-Brahim, A. Gastli, and K. Ghazi, "An improved predictive control approach for Multilevel Inverters," in *Sensorless Control for Electrical Drives and Predictive Control of Electrical Drives and Power Electronics (SLED/PRECEDE), 2013 IEEE International Symposium on*, 2013, pp. 1–7.
43. M. Ghanes, M. Trabelsi, H. Abu-Rub, and L. Ben-Brahim, "Robust adaptive observer-based model predictive control for multilevel flying capacitors inverter," *IEEE Trans. Ind. Electron.*, vol. 63, no. 12, pp. 7876–7886, 2016.
44. M. Trabelsi, L. Ben-Brahim, and K. Ghazi, "An improved real-time digital feedback control for grid-tie multilevel inverter," in *IECON 2013-39th Annual Conference of the IEEE Industrial Electronics Society*, 2013, pp. 5776–5781.
45. M. Trabelsi, K. Ghazi, N. Al-Emadi, and L. Ben-Brahim, "An original controller design for a grid connected PV system," in *IECON 2012-38th Annual Conference on IEEE Industrial Electronics Society*, 2012, pp. 924–929.
46. L. Ben-Brahim, A. Gastli, M. Trabelsi, K. A. Ghazi, M. Houchati, and H. Abu-Rub, "Modular multilevel converter circulating current reduction using model predictive control," *IEEE Trans. Ind. Electron.*, vol. 63, no. 6, pp. 3857–3866, 2016.

Chapter 4
Packed U-Cell Topology

Packed U-Cells (PUC) inverter is an emerging single-DC-source MLI topology, which facilitates implementing existing controllers for various applications. It is a competitive topology combining multiple features of other MLI topologies [1] such as: (1) low impact on the power grid; (2) flexibility in expanding to higher output levels without DC bus extension; (3) ability to offer extended selection of control actions and improve the filters bandwidth through the switching states redundancy; (4) reliability and low cost due to the reduced number of active components; (5) better ride through capability by means of the existing storage capacitors.

4.1 Configuration

Different drawings of the PUC inverter have been shown in Fig. 4.1 [2, 3]. It has only 6 active switches and 2 DC buses. it is a Single-DC-Source topology while the lower DC link capacitor (C_2) is a flying capacitor and its voltage is controlled through switching algorithm or external controllers. It has the lower number of components between other topologies for the same number of generated voltage levels. Each cell of PUC inverter consists of one capacitor and two switching devices. Considering n cells, the inverter will consist of $2n$ switches (the two switches of the same cell must be controlled in a complementary way, which will give $2n$ combinations with redundant states) and n-1 capacitors. Table 4.1 shows the different switching states for the 3-cell configuration.

According to Table 4.1, if the capacitor C_2 voltage is controlled at 1/3 of V_{dc}, then a 7-level voltage waveform would be generated at the output terminals. The 7-level PUC inverter has been analyzed in [4] however it requires a very complicated controller to balance the flying capacitor voltage [5–7]. Various linear/nonlinear voltage controllers have been designed and implemented on this topology to reach a 7-level

© The Author(s), under exclusive license to Springer Nature Switzerland AG 2019
H. Vahedi, M. Trabelsi, *Single-DC-Source Multilevel Inverters*,
SpringerBriefs in Electrical and Computer Engineering,
https://doi.org/10.1007/978-3-030-15253-6_4

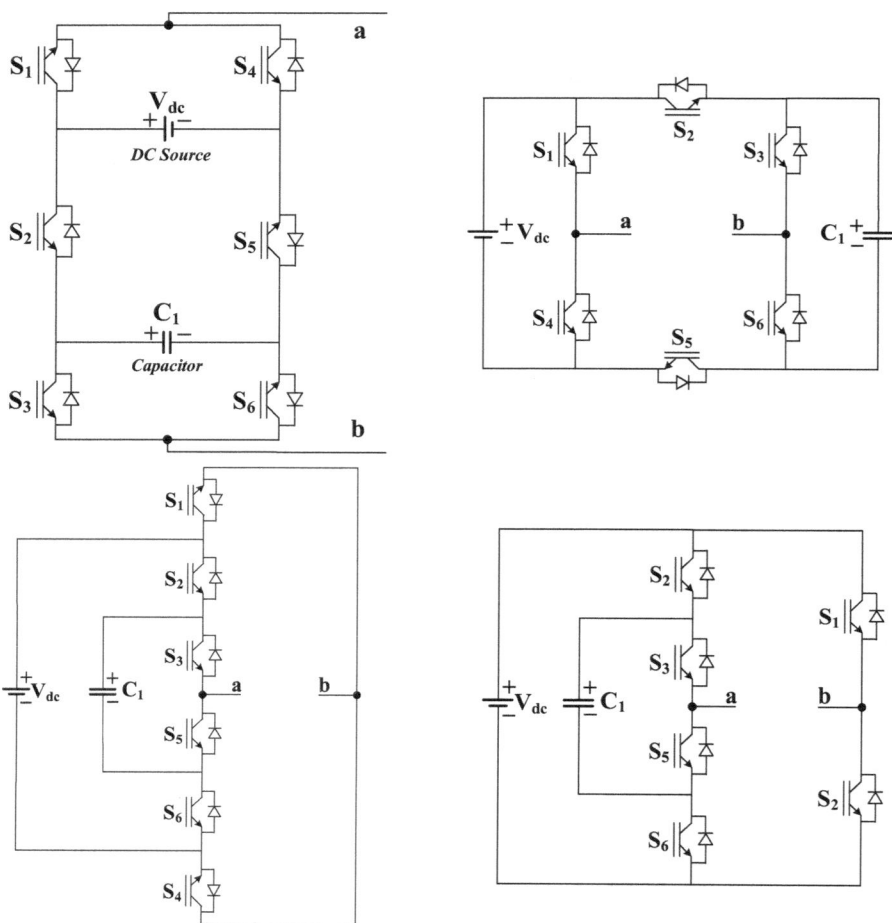

Fig. 4.1 PUC inverter configuration in different drawings

Table 4.1 All possible switching states of the PUC inverter

State	V_{ab}	s_1	s_2	s_3
1	V_{dc}	1	0	0
2	$V_{dc}-V_{C1}$	1	0	1
3	V_{C1}	1	1	0
4	0	1	1	1
5	0	0	0	0
6	$-V_{C1}$	0	0	1
7	$V_{C1}-V_{dc}-$	0	1	0
8	$-V_{dc}$	0	1	1

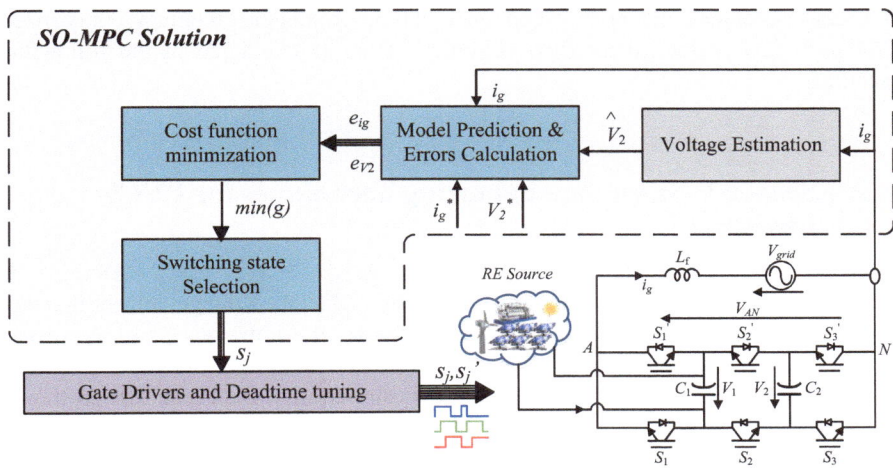

Fig. 4.2 Synoptic of the proposed SO-MPC solution for a grid-connected PUC7 inverter

Fig. 4.3 Estimated
capacitor voltage compared
to the measurement and the
reference values

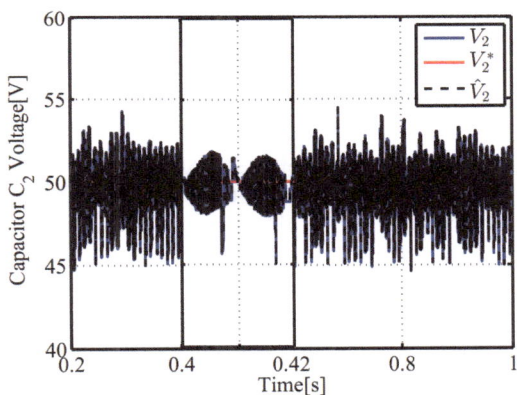

waveform at the output [8]. For instance, in [9], authors presented a Switched
Observer (SO) for reduced-sensor control of grid-connected PUC7 inverter. Using
the measured grid current, the proposed SO estimates the PUC capacitor voltage to
be fed to the control algorithm (Figs. 4.2 and 4.3).

Noticing the switching states, it can be revealed that controlling the capacitor
voltage at ½ of Vdc leads to generate a 5-level voltage waveform at the output.
Using the same topology arrangement, the 5-level version is considered as a modification of the original 7-level configuration with less voltage levels but higher reliability and controllability [10]. The 5-level PUC inverter is called PUC5 that has
found industrial applications thanks to simplicity, high stability and reliably of voltage balancing algorithm. As a single-dc-source MLI topology, the V_{dc} can be
replaced by a PV panel or battery and its associated controller would be same as all

standard cascaded controllers as explained in [11, 12]. On the other hand, the capacitor voltage is balanced through switching states as explained in the following section.

4.2 Sensor-Less Voltage Balancing Techniques for PUC5 Inverter

The use of DC capacitor in the PUC5 topology makes the voltage control mandatory as for any other MLI topology. Motivated by the importance of the correct knowledge of the capacitors' voltages in the control design (Table 4.2), recent research works were focused on the accurate estimation of the capacitors' voltages for different PUC5 configurations [13–16]. The proposed sensor-less voltage technique reduces the complexity of the control system, which makes the PUC5 inverter appealing for industrial applications.

In [10], a sensor-less voltage control technique based on redundant switching states was designed for PUC5 inverter. The proposed technique regulates the dc capacitor voltage at half of the dc source value allowing the generation of a 5-level output voltage waveform with low harmonic distortion. The PUC auxiliary dc bus is regulated only by sensor-less controller integrated into the modulation technique as depicted in Fig. 4.4. Cr_1 to Cr_4 are the standard carriers in a 5-level PWM scheme, which are shifted vertically to modulate the reference signal of V_{ref}.

As analyzed in Table 4.2, states 2 and 6 are used in PUC5 inverter in order to balance the capacitor voltage at the desired level. Those two configurations are shown in Fig. 4.5. A line inductor is used to derive the relationship between V_1 and V_2 in steady-state operation as shown by Eqs. (4.1)–(4.10).

The capacitor current should be used to analyze its voltage balancing principle. As seen from the above figure, the current i_c equals to i_s during the switching states in which the capacitor is involved. Therefore, the following equations could be written for that current:

$$i_c = i_s \qquad\qquad (4.1)$$

Table 4.2 Capacitors voltages variations for the different switching patterns

S_1	S_2	S_3	C_2
1	0	0	Bypassed
1	0	1	Charged
1	1	0	Discharged
1	1	1	Bypassed
0	0	0	Bypassed
0	0	1	Discharged
0	1	0	Charged
0	1	1	Bypassed

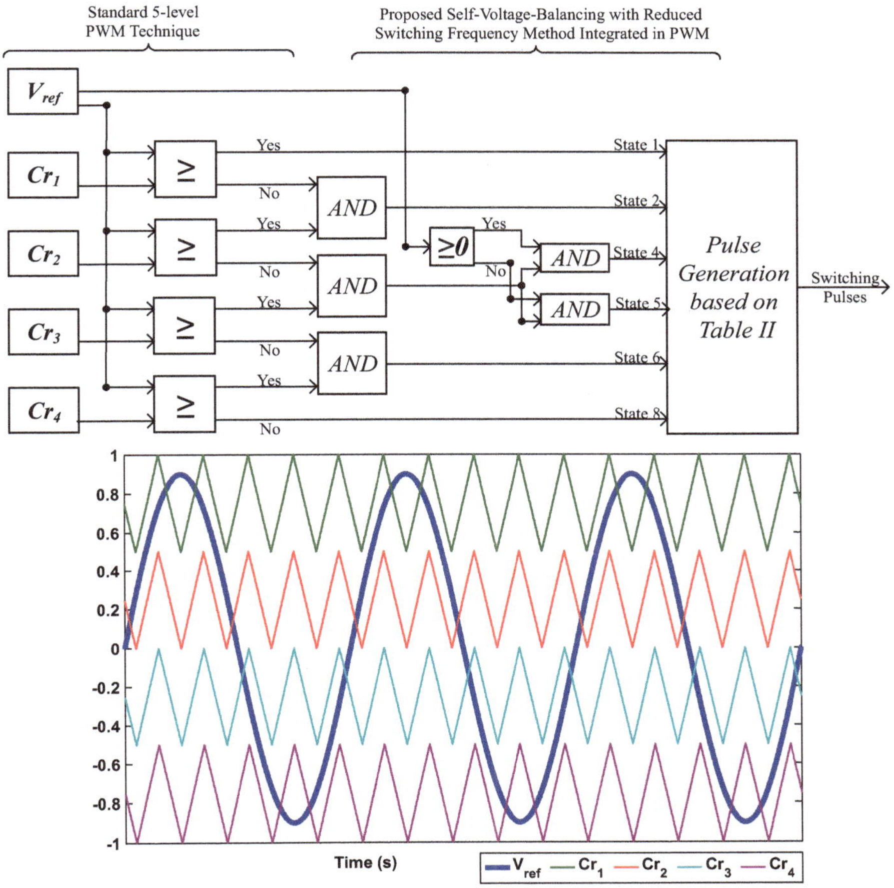

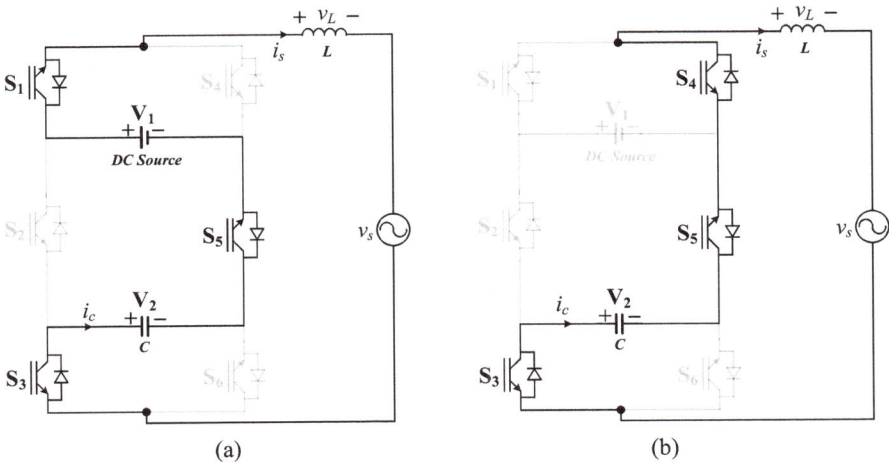

Fig. 4.4 Proposed open-loop switching algorithm for sensor-less self-voltage balancing of PUC5 inverter

Fig. 4.5 PUC5 configuration during (**a**) charging, and (**b**) discharging

$$v_L = L\frac{di_s}{dt} \tag{4.2}$$

The capacitor charge balance can be written as:

$$\int_{char.} i_s\,dt + \int_{dischar.} i_s\,dt = 0 \tag{4.3}$$

The capacitor voltage V_2 is assumed ripple free.

For instance, assume a charging interval of duration t_p and a discharging interval of duration t_n typically as shown in Fig. 4.6.

During the charging interval, Fig. 4.5a and Eq. (4.2) give

$$\begin{aligned}
i_s &= \frac{1}{L}\int_t^o (V_1 - V_2 - v_s)\,dt + i_{so} \\
&= \frac{V_1 - V_2}{L}\int_t^o dt - \frac{1}{L}\int_t^o v_s\,dt + i_{so}
\end{aligned} \tag{4.4}$$

Leading to

$$i_s = \left(\frac{V_1 - V_2}{L}\right)t + i_{so} - \int_t^o v_s\,dt \tag{4.5}$$

During the discharging interval, and assuming the time origin is now at the beginning of this interval, Fig. 4.15b and Eq. (4.2) give

$$\begin{aligned}
i_{s'} &= \frac{1}{L}\int_{t'}^{o'} (-V_2 - v_s)\,dt' + i_{so'} \\
&= \left(-\frac{V_2}{L}\right)t' + i_{so'} - \frac{1}{L}\int_{t'}^{o'} v_s\,dt'
\end{aligned} \tag{4.6}$$

Where, i_{so} and $i_{so'}$ are the initial currents of charging and discharging intervals, respectively. Applying the charge balance as in Eq. (4.3), yields

Fig. 4.6 Output 5-level voltage waveform and typical charging/ discharging intervals

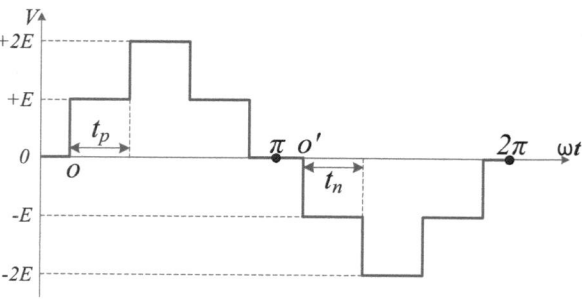

$$\overset{o}{\underset{t_p}{\int}} i_s \, dt + \overset{o}{\underset{t_n}{\int}} i_{s'} \, dt' = 0 \tag{4.7}$$

$$\frac{V_1 - V_2}{2L} t_p^2 - \frac{V_2}{2L} t_n^2 + i_{so} t_p + i_{so'} t_n$$
$$- \frac{1}{L} \left[\overset{o}{\underset{t_p}{\int}} \left(\overset{o}{\underset{t}{\int}} v_s \, dx \right) dt + \overset{o}{\underset{t_n}{\int}} \left(\overset{o'}{\underset{t'}{\int}} v_s \, dx \right) dt' \right] = 0 \tag{4.8}$$

The sinusoidal shapes of current i_s and voltage v_s, as well as symmetry in the control processed error (as shown below), imply that

$$i_{so} = -i_{so'} \ \text{ and } \ \overset{o}{\underset{t_p}{\int}} \left(\overset{o}{\underset{t}{\int}} v_s \, dx \right) dt + \overset{o}{\underset{t_n}{\int}} \left(\overset{o'}{\underset{t'}{\int}} v_s \, dx \right) dt' = 0 \tag{4.9}$$

Thus, the charge balance expression simplifies to

$$\frac{V_1 - V_2}{2L} t_p^2 - \frac{V_2}{2L} t_n^2 + i_{so} \left(t_p - t_n \right) = 0 \tag{4.10}$$

Furthermore, it will be shown that, owing to the half-wave symmetry in the control signal, to every charging duration t_p corresponds an equal discharging time duration t_n.

As illustrated in Fig. 4.4, Since the reference waveform (V_{ref}) which is sent to the modulator is a symmetric one, the charging and discharging time would be equal (due to modulating by fixed frequency and symmetric carriers). However, the following relations prove the symmetrical shape of the V_{ref}.

V_{ref} is a sine wave in stand-alone mode of operation imposed as open-loop system input. Therefore, it has naturally a symmetric shape without requiring any proof and the output pulses of the modulator would have half wave symmetry shape with equal timing on the switching states of each half cycle.

For the closed-loop system, as grid-connected mode of operation, it should be demonstrated that the output of control unit is symmetric which is assumed as V_{ref}. for instance, in a current control loop, the input of the PI controller is the error signal defined as Eq. (4.11). The processing expression of a PI controller in time domain is written as Eq. (4.12).

$$e = i^* - i$$
$$= a \sin(\omega t) - b \sin(\omega t)$$
$$= (a - b) \sin(\omega t) \tag{4.11}$$
$$= c \sin(\omega t)$$

$$G(PI) = k_p \, e + k_i \int e \, dt \tag{4.12}$$

By calculating the PI output with the given input of *e*, the following relation is attained:

$$output\ of\ PI = k_p c \sin(\omega t) + k_i \int c \sin(\omega t) dt$$

$$= k_p c \sin(\omega t) - \frac{k_i c}{\omega} \cos(\omega t) \qquad (4.13)$$

$$= K \sin(\omega t - \varphi)$$

The calculated output is still a sine wave that ensures the half wave symmetry of the V_{ref}.

Considering the above effort in proving the fact the V_{ref} has half wave symmetry as sine wave, it is then obvious that the switching states 2 and 6 would have equal intervals. Going back to Eq. (4.10), the following relation is achieved:

$$V_1 = 2V_2 \qquad (4.14)$$

So, it can be concluded that the capacitor voltage tracks the half of DC source amplitude acceptably through the charging and discharging switching states (2 and 6) by applying the proposed switching technique illustrated in Fig. 4.4.

4.3 Design Criteria

One important design consideration of power converters in industries is the use of identical devices in the product. For multilevel inverters, it is so much challenging to have equal voltage rating switches as well as using switch modules instead of single switch.

Analyzing the PUC5 configuration gives the detailed information of switches voltages ratings [1]. Assuming that $V_1 = 2E$ in PUC5 inverter, it is observable from Fig. 4.1 that two upper switches S_1 and S_2 should suffer $2E$ and the other 4 switches have equal voltage rating of E. first idea to design the PUC5 inverter could be using of identical voltage rating switches. Hence, two switches could be added to the upper ones in series to have equal voltage ratings. As shown in Fig. 4.7a, 4 half-bridge modules as shaded boxes would be used to form the PUC5 configuration.

Another idea to design the PUC5 inverters is based on the fundamental switching frequency of S_1 and S_2. As illustrated in Fig. 4.8, the two upper switches that have higher voltage ratings, they turn ON once in a cycle with the line frequency. So, the lowest working frequency switch in the market could be used there. The four other switches work at same switching frequency and voltage rating. Therefore, according to the Fig. 4.7b, 3 half-bridge modules are connected to form the PUC5.

The final consideration about switches in the PUC5 inverter structure is the fact that they are working complimentarily. So, 3 main pulses drive S_1, S_2 and S_3 while their NOT are sent to the S_4, S_5 and S_6, respectively. Moreover, it should be noted that S_4 and S_5 are common emitter that means they need one source for the gate drivers.

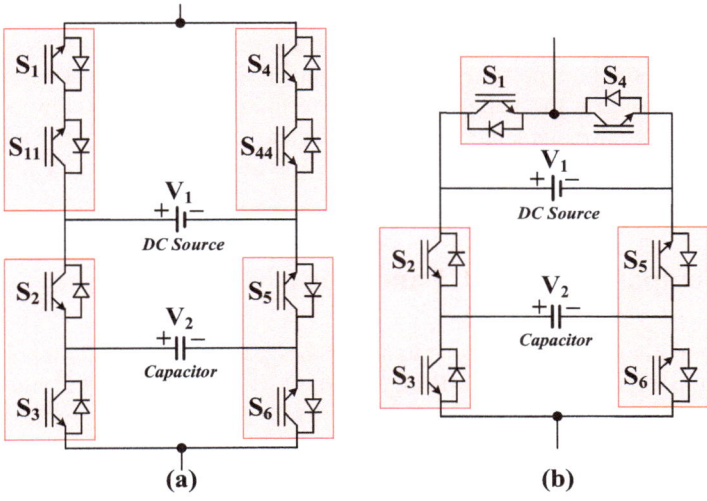

Fig. 4.7 PUC5 inverter design with (**a**) equal switch voltage rating (**b**) with low frequency module

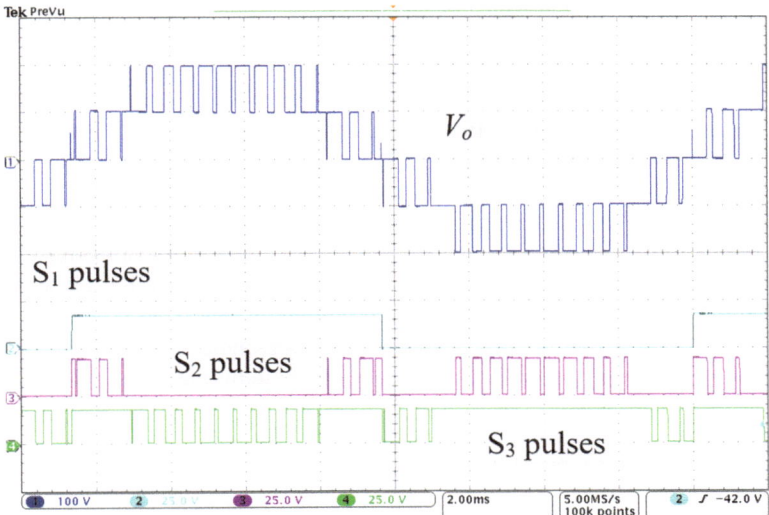

Fig. 4.8 PUC5 inverter switching pulses

According to the PUC5 modelling performed in [10], all switches should have equal current rating same as the output current (i_o). The capacitor voltage should be equal to the half of V_1 however in case of any failure, its voltage should be considered equal to the V_1 as maximum DC voltage in the circuit. It should be mentioned that the capacitor voltage can be controlled with the above-mentioned sensor-less technique or through redundant switching states as a conventional method of voltage balancing in multilevel inverters with flying capacitors. The latter needs to use all switching states listed in Table 4.2. Which results in using a very small size flying capacitor at high power ratings. The voltage ripple of the capacitor should be considered based on the voltage balancing technique. If the capacitor voltage is regulated using all redundant switching states and a voltage feedback sensor, then the capacitance could be reduced to less than 100 μF for a 3 kW 200 V rating.

4.4 Comparison of PUC5 and Other MLIs topologies

Some comparisons should be done to assess the excellence of PUC5 topology. The selected topologies include Full-Bridge (FB), CHB, NPC, T3, ANPC and FC [17]. The idea of selecting those topologies was that they are the most popular ones and manufactured by industries. Moreover, except CHB, they can be single-DC-source topologies. Additionally, all of them except FB, generate 5-level voltage waveform at the output. Too many topologies have been reported as multiple-DC-source, which are not interesting for the industries anymore. The comparison summary has been listed in Table 4.3.

The above table can be analyzed by each column. Noticing the number of levels, FB is removed from comparison since it is a 3-level topology. The higher number of levels, the lower THD, the smaller size of filter and manufactured product. The next column is the number of isolated DC sources that means a bulky transformer plus a diode bridge or PV panel or batteries. Their prices are much more than switches so CHB is eliminated in the comparison process. Considering the next three columns, PUC5 has lower number of components among other topologies of NPC, T3, FC and ANPC. Moreover, by defining the component to level ratio, it is still distinguishing that the PUC5 structure generates more voltage levels while using less components. Eventually, as explained in previous sections, the voltage balancing of the auxiliary capacitor in the PUC5 inverter is performed through the redundant switching states without requiring any external controller. That means the PUC5 inverter, unlike the other topologies, does not impose any complexity to the control design process. A conventional cascaded controller with one voltage loop and one

Table 4.3 Results of comparison between PUC5 and reported topologies

Topology	Levels	DC Source	Capacitor	Switch	Diode	Component to level ratio	Voltage balancing
FB	3	1	0	4	0	1.66	No Need
CHB	5	2	0	8	0	2	External Regulator
NPC	5	1	2	8	4	3	External Regulator
T3	5	1	2	8	0	2.2	External Regulator
FC	5	1	2	8	0	2.2	Redundant States
ANPC	5	1	3	8	0	2.4	External Regulator
PUC5	5	1	1	6	0	1.6	Redundant States

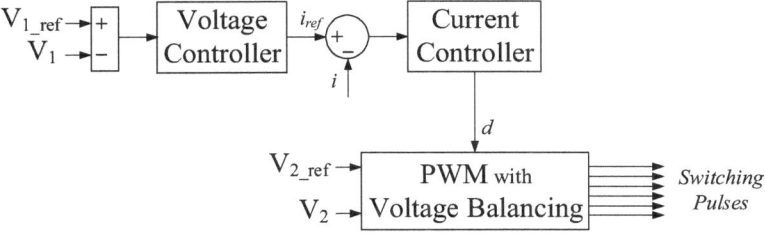

Fig. 4.9 Conventional cascaded controller applied on the PUC5 converter

current loop can be implemented on the PUC5 converter in various applications such as grid-connected inverter for renewable energy conversion system, drive, active filter, rectifier, UPS, etc. as shown in Fig. 4.9. The voltage controller block could be a PI to regulate the upper DC link (V_1). Then the reference current (i_{ref}) is compared to the actual line current (i) and the error is minimized through the current controller which can be type of linear or nonlinear ones like PI, PR, model predictive, back stepping, sliding mode, etc. the generated signal d represents the duty cycle that is modulated by the PWM block. However, the switching technique is manipulated to receive the voltage feedback from auxiliary capacitor; compare it with reference value that is $V_1/2$, and choose the appropriate pulses between redundant states in order to keep the V_2 balanced. Unlike the controllers for multiple-DC-source converters as shown in Fig. 4.6, the implementable controller on PUC5 does not require any additional voltage regulator to take care of auxiliary capacitor voltage. It is exactly same as conventional cascaded controllers used in industries with slight modification in the modulation block. It is certainly the most amazing feature of single-DC-source PUC5 converter in which the capacitor voltage is balanced through redundant switching states.

4.5 PUC5 Inverter Applications

4.5.1 Standalone Inverter

In this case, the output voltage of the PUC inverter is assumed to be a 5-level waveform while the capacitor voltage (V_2) is regulated automatically without using any voltage sensor. This sensor-less controller is integrated into switching technique using redundant switching sequences and works even in low switching frequency. Figure 4.10 illustrates the test results of a sensor-less PUC5 inverter in standalone mode supplying RL (a resistor and an inductor in series) load and adding nonlinear load while V_2 is regulated at half of the V_1. Such operation can be used for UPS, motor drive or Vehicle-to-Home (V2H) applications in which the battery is connected to the DC bus and loads are supplied by a 5-level low harmonic voltage on the AC side.

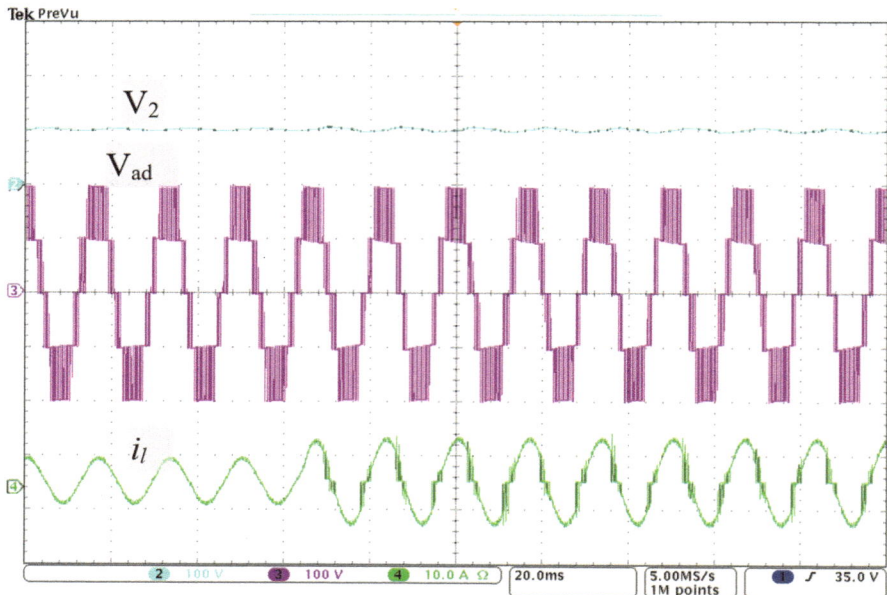

Fig. 4.10 Adding single-phase rectifier (as nonlinear load) paralleled with the RL load to the output of Sensor-less PUC5 standalone inverter

4.5.2   Grid-Connected Inverter

The grid-connected operation has been tested on the sensor-less PUC5 inverter and results are depicted in Fig. 4.11. The unity power factor operation is ensured by a PI controller sensing only the grid side voltage and current. The PUC5 capacitor voltage is still regulated without any sensor. The high resolution and sharp 5-level voltage waveform validate the good performance of the proposed converter even in high dynamic grid-connected applications. Such result proves that the PUC5 inverter can be used in renewable energy conversion system as PV inverter or Vehicle-to-Grid (V2G) applications.

4.5.3   PUC5 Rectifier

The practical results of the PUC5 rectifier have been shown in Fig. 4.12 and prove the good dynamic performance of the controller and voltage balancing technique which has been integrated into the modulation process. The auxiliary capacitor voltage is kept regulated at desired level with low voltage ripple due to such fast and accurate voltage balancing approach results in generating a 5-level quasi-sine voltage waveform at the input of the rectifier with low harmonic contents. Such

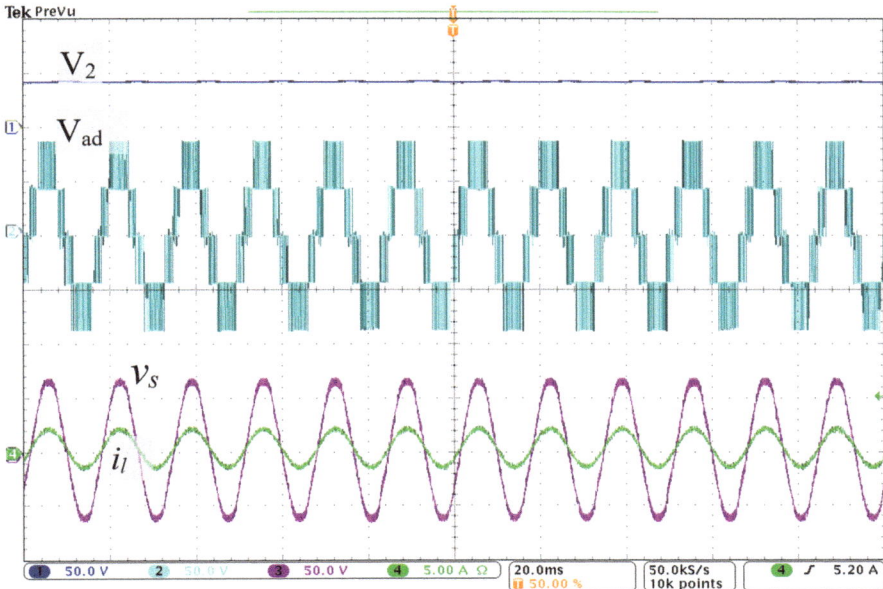

Fig. 4.11 Sensor-less PUC5 grid-connected inverter voltage and current waveforms

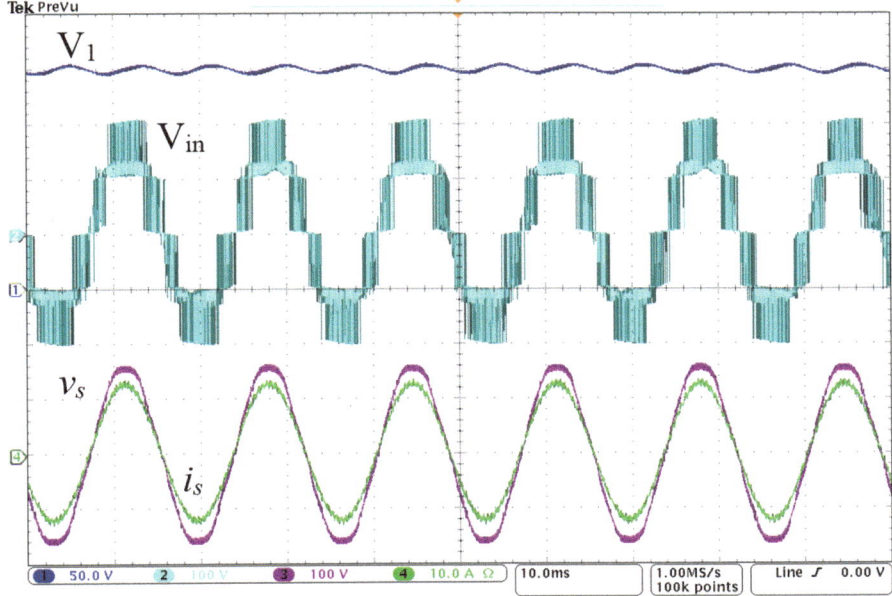

Fig. 4.12 Experimental results of PUC5 rectifier

multilevel waveform helps using small size filter to eliminate the line current harmonics. Moreover, the unity power factor mode of operation is achieved by the standard cascaded controller as shown in Fig. 4.9. The PUC5 rectifier could be a potential candidate to be used as industrial rectifier in traction systems or battery charger for EV applications.

4.5.4   PUC5 STATCOM

Night shift inverters are now getting attention from industries to inject active power from PV panels at daytimes and provide reactive power for the grid at nights. Using multilevel converters can reduce harmonic contents, switching frequency and power losses significantly. The PUC5 STATCOM has been analyzed in this application and a suitable controller has been designed to inject reactive power into the grid while regulating the DC links voltages without using voltage sensors. Results are shown in Fig. 4.13. 90° phase shift between grid voltage and current ensures the high efficiency operation of the PUC5 converter in injecting reactive power into the grid.

In this application, same as rectifier mode of operation, the PUC5 converter is connected to the grid and the DC links voltage should be regulated at desired level. However, instead of connecting any loads to the DC bus and correcting the input power factor, the converter will inject a controlled amount of reactive power to the grid as a variable capacitor which is ensured by 90° phase shift between grid side

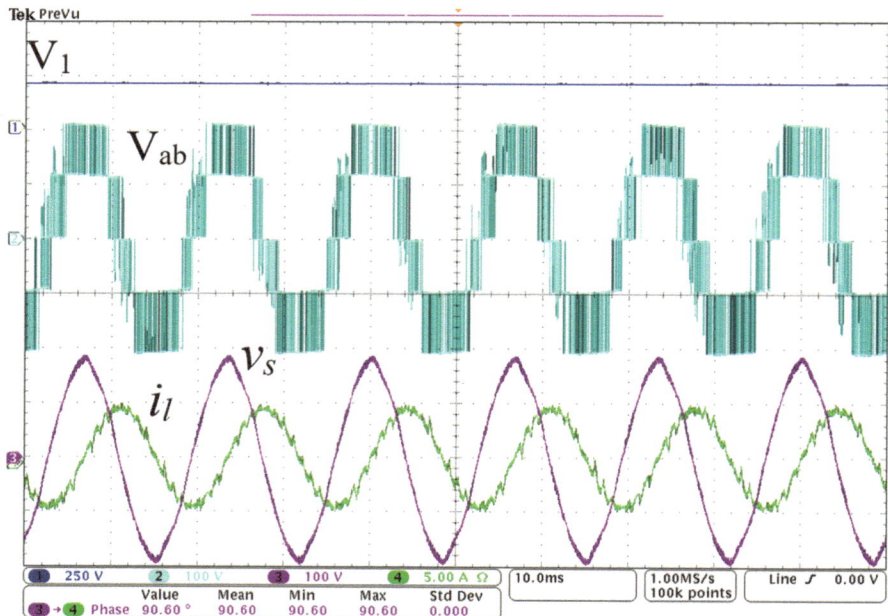

Fig. 4.13 Experimental results of PUC5 STATCOM

voltage and current waveforms. This topology can replace the installed bulky capacitors in power system to compensate the required reactive power at any level.

4.5.5 PUC5 Dynamic Voltage Restorer

Power quality is the major concern of all power system participants. It can appear in different aspects such as voltage sag or swell, long interruptions, harmonic components, etc. Conventionally, voltage sags and swells are considered as disturbances but not failures or interruptions. Dynamic voltage restorer (DVR) is a power electronics equipment that can solve the aforementioned problems in a power system with voltage disturbance issues. It is an active inverter that injects a compensation voltage in series and synchronized with the grid at the point of common coupling. Thus, taking advantage of its high ride through capability, a PUC5 inverter-based DVR was proposed in [18]. In the control structure, an MPC strategy was implemented to control the system variables. A discrete-time model of the whole system was used to predict at each sampling time the filter current, the ancillary capacitor voltage, and the compensation voltage considering all switching patterns (Fig. 4.14).

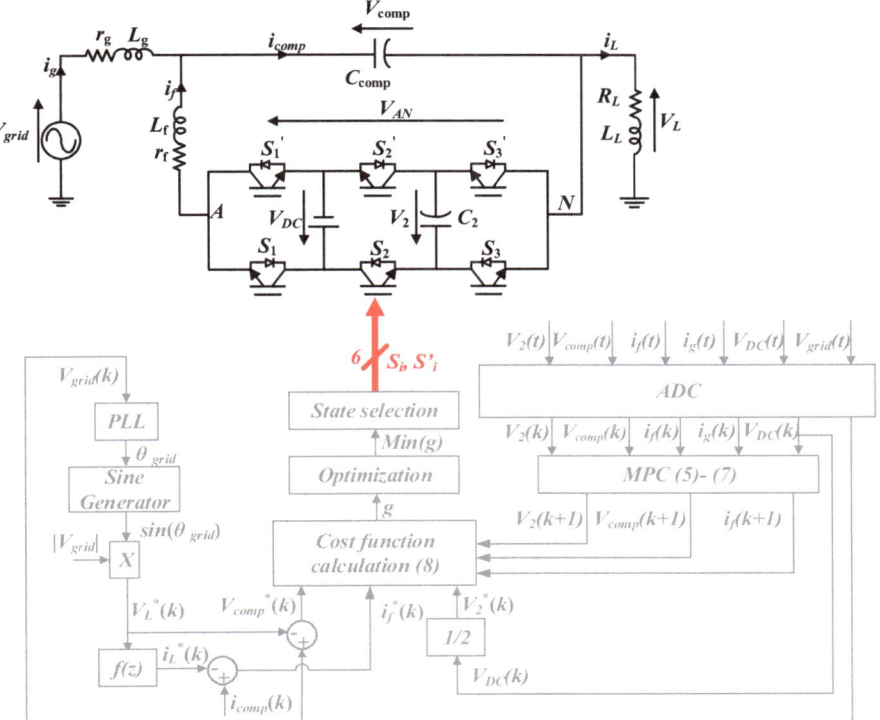

Fig. 4.14 Block diagram of the MPC controlled PUC5 based DVR

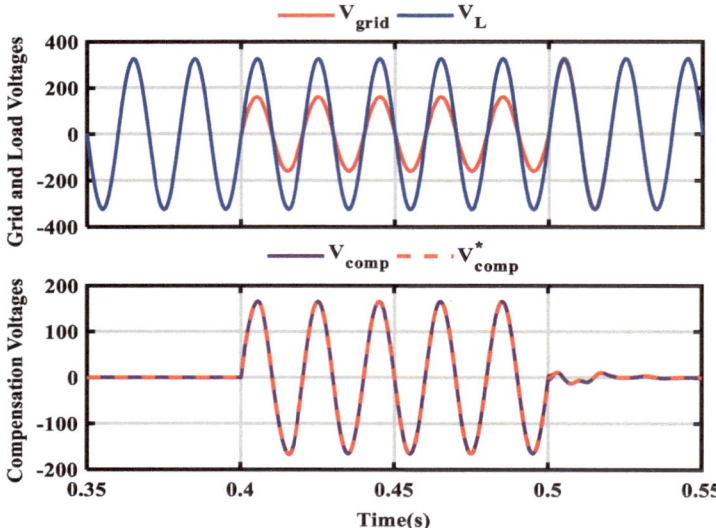

Fig. 4.15 Dynamic responses of the grid voltage V_{grid}, the load voltage V_L, and the compensation voltage V_{comp} during grid voltage sag

Figure 4.15 depicts the simulation results during short-term decrease of the RMS grid voltage (voltage sag during 5 cycles), where the grid voltage, load voltage, and compensation voltage are displayed. One can note that the generated compensation voltage is maintained around zero before the grid voltage sag. At the voltage sag event, the required compensation voltage is produced by the DVR in order to keep the sensitive load voltage unaffected and maintained at the desired level, which confirms the good performance of the proposed solution.

4.5.6   PUC5 Three-Phase Inverter

The three-phase inverters can be used in motor drives, UPS and traction system to deliver high power to the loads. The higher levels, the lower THD, the lower output filter size. Therefore, the PUC5 has been tested as three-phase inverter by connecting three single-phase PUC5 units as shown in Fig. 4.16. It can be connected in 3-wire or 4-wire configurations. One of the main advantages of PUC5 inverter against other multilevel inverters is using only one isolated DC source so in three-phase application only three isolated DC sources are required. Figure 4.17 illustrates the voltages waveforms of a three-phase PUC5 inverter supplying an RL load.

As seen from Fig. 4.17, the phase voltage, line voltage and load voltage have 5, 9 and 14 levels, so the THD would be so much less than conventional 2-level converters.

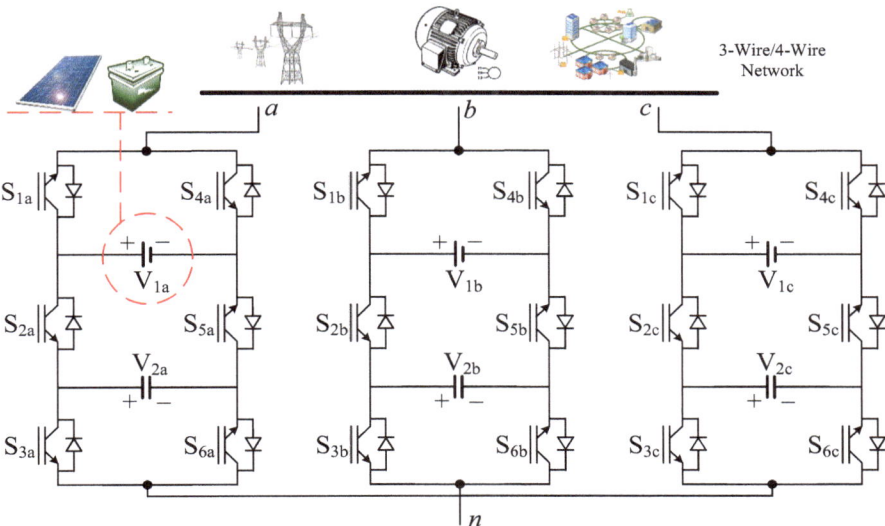

Fig. 4.16 Three-phase PUC5 inverter configuration

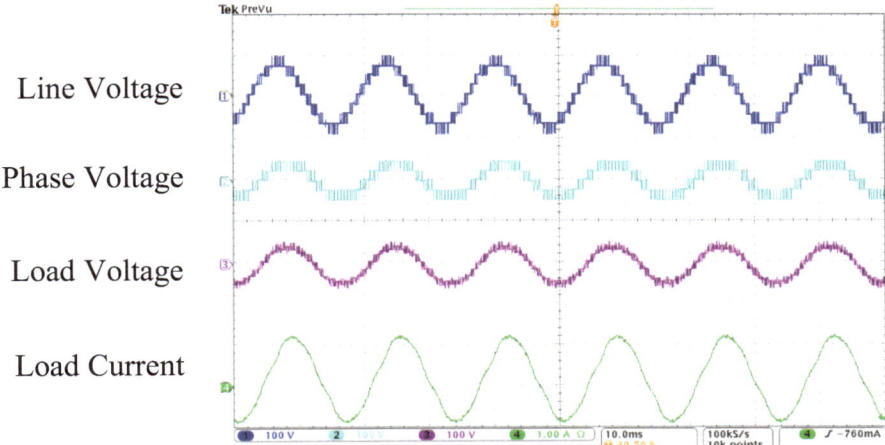

Fig. 4.17 Three-phase PUC5 inverter voltage and current waveforms

The PUC5 inverter has been analyzed based on the topology and control design. It has been compared to some mostly used topologies to reveal the promising features of that configuration. The main advantage of PUC5 inverter is the redundant switching states, which eliminate the need for external voltage regulator of auxiliary capacitor. The sensor-less voltage balancing technique applicable on the PUC5 inverter has been explained theoretically. It was shown that the auxiliary capacitor could be controlled without using any voltage feedback surprisingly.

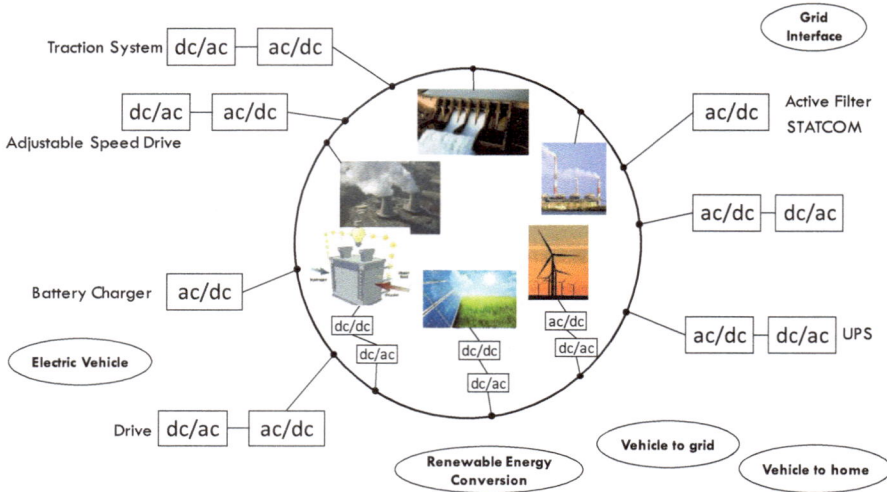

Fig. 4.18 Possible applications of PUC5 inverter

Moreover, the detailed analysis of single-DC-source and multiple-DC-source topologies along with the associated controller designs indicated the fact that single-DC-source converter are much more interesting for power industries due to lower manufacturing price and less complicated controllers. It has been demonstrated that the conventional cascaded controller can be simply applied on the PUC5 converter without additional regulators since the auxiliary capacitor voltage is regulated through redundant switching states acceptably. Eventually, the full assessment among FB, CHB, NPC, T3, FC, ANPC and PUC5 topologies proved that the presented configuration is a potential candidate to compete in the market of power electronics converters and full range of applications is expected for PUC5 inverter as shown in Fig. 4.18.

It should be mentioned that the PUC5 technology has been licensed to Ossiaco Inc., Montreal for commercialization.

References

1. H. Vahedi and K. Al-Haddad, "PUC5 Inverter–A Promising Topology for Single-Phase and Three-Phase Applications," in *IECON 2016-42nd Annual Conference of the IEEE Industrial Electronics Society*, Italy, 2016.
2. K. Al-Haddad, Y. Ounejjar, and L. A. Gregoire, "Multilevel Electric Power Converter," US Patent 20110280052, 2011.
3. H. Vahedi and K. Al-Haddad, "METHOD AND SYSTEM FOR OPERATING A MULTILEVEL INVERTER," US Patent US9923484B2, 2018.
4. Y. Ounejjar, K. Al-Haddad, and L. A. Grégoire, "Packed U cells multilevel converter topology: theoretical study and experimental validation," *IEEE Trans. Ind. Electron.*, vol. 58, no. 4, pp. 1294–1306, 2011.

5. Y. Ounejjar, K. Al-Haddad, and L. A. Dessaint, "A Novel Six-Band Hysteresis Control for the Packed U Cells Seven-Level Converter: Experimental Validation," *IEEE Trans. Ind. Electron.,* vol. 59, no. 10, pp. 3808–3816, 2012.

6. H. Vahedi and K. Al-Haddad, "Real-Time Implementation of a Seven-Level Packed U-Cell Inverter with a Low-Switching-Frequency Voltage Regulator," *IEEE Trans. Power Electron.,* vol. 31, no. 8, pp. 5967–5973, 2016.

7. J. Metri, H. Vahedi, H. Kanaan, and K. Al-Haddad, "Real-Time Implementation of Model Predictive Control on 7-Level Packed U-Cell Inverter," *IEEE Trans. Ind. Electron.,* vol. 63, no. 7, pp. 4180–4186, 2016.

8. M. Trabelsi, M. Ghanes, M. Mansouri, S. Bayhan, and H. Abu-Rub, "An original observer design for reduced sensor control of Packed U Cells based renewable energy system," *International Journal of Hydrogen Energy,* vol. 42, no. 28, pp. 17910–17916, 2017.

9. M. Trabelsi, M. Ghanes, S. Bayhan, and H. Abu-Rub, "High performance voltage-sensorless model predictive control for grid integration of packed U ceils based PV system," in *19th European Conference on Power Electronics and Applications (EPE'17 ECCE Europe),* 2017, pp. P. 1-P. 8.

10. H. Vahedi, P. Labbe, and K. Al-Haddad, "Sensor-Less Five-Level Packed U-Cell (PUC5) Inverter Operating in Stand-Alone and Grid-Connected Modes," *IEEE Trans. Ind. Informat.,* vol. 12, no. 1, pp. 361–370, 2016.

11. H. Vahedi, A. Shojaei, A. Chandra, and K. Al-Haddad, "Five-Level Reduced-Switch-Count Boost PFC Rectifier with Multicarrier PWM," *IEEE Trans. Ind. Applications,* vol. 52, no. 5, pp. 4201–4207, 2016.

12. H. Vahedi, A. Shojaei, L.-A. Dessaint, and K. Al-Haddad, "Reduced DC Link Voltage Active Power Filter Using Modified PUC5 Converter," *IEEE Trans. Power Electron.,* vol. 33, no. 2, pp. 943–947, 2018.

13. M. Abarzadeh, H. Vahedi, and K. Al-Haddad, "Fast Sensor-Less Voltage Balancing and Capacitor Size Reduction in PUC5 Converter Using Novel Modulation Method," *IEEE Trans. Ind. Informat.,* vol. Early Access, no. PP, p. 1, 2019.

14. H. Vahedi, M. Sharifzadeh, and K. Al-Haddad, "Topology and control analysis of single-DC-source five-level packed U-cell inverter (PUC5)," in *IECON 2017-43rd Annual Conference of the IEEE Industrial Electronics Society,* 2017, pp. 8691–8696.

15. S. Arazm, H. Vahedi, and K. Al-Haddad, "Phase-shift modulation technique for 5-level packed U-cell (PUC5) inverter," in *IEEE 12th International Conference on Compatibility, Power Electronics and Power Engineering (CPE-POWERENG),* 2018, pp. 1–6.

16. S. Arazm, H. Vahedi, and K. Al-Haddad, "Space vector modulation technique on single phase sensor-less PUC5 inverter and voltage balancing at flying capacitor," in *IECON 2018-44th Annual Conference of the IEEE Industrial Electronics Society,* 2018, pp. 4504–4509.

17. L. G. Franquelo, J. Rodriguez, J. I. Leon, S. Kouro, R. Portillo, and M. A. M. Prats, "The age of multilevel converters arrives," *IEEE Ind. Electron. Mag.,* vol. 2, no. 2, pp. 28–39, 2008.

18. M. Trabelsi, H. Vahedi, H. Komurcugil, H. Abu-Rub, and K. Al-Haddad, "Low complexity model predictive control of PUC5 based dynamic voltage restorer," in *ISIE2018-27th International Symposium on Industrial Electronics,* 2018, pp. 240–245.